Where Is the Sun?

Ready-to-Read

Simon Spotlight
New York London Toronto Sydney New Delhi

Here is a list of all the words you will find in this book. You might like to sound them out before you begin reading the story.

Names:

Plants

Rain

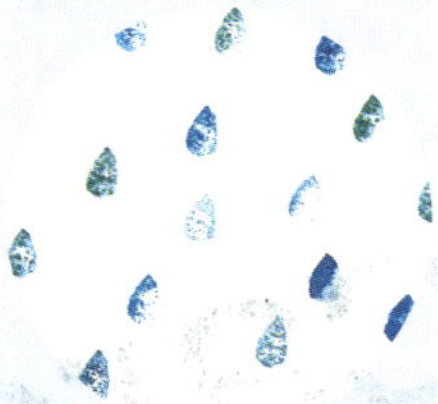

Sun

SIMON SPOTLIGHT
An imprint of Simon & Schuster Children's Publishing Division • New York London Toronto Sydney New Delhi • 1230 Avenue of the Americas, New York, New York 10020 • This Simon Spotlight edition December 2019 •
For information about special discounts for bulk purchases, please contact Simon & Schuster Special Sales at 1-866-506-1949 or business@simonandschuster.com.
Manufactured in the United States of America 1119 LAK • 10 9 8 7 6 5 4 3 2 1
Library of Congress Cataloging-in-Publication Data • Names: Carle, Eric, author, illustrator. Title: Where is the sun? / Eric Carle. Description: New York, New York : Simon Spotlight, an imprint of Simon & Schuster Children's Publishing Division, 2019. | Audience: Grades K–1. | Summary: "Where is the sun? Beginning readers will love learning to read and love learning about the sun's position in the sky, and the opposites between night and day"— Provided by publisher. Identifiers: LCCN 2019024264 | ISBN 9781534455474 (paperback) ISBN 9781534455481 (hardcover) Subjects: LCSH: Sun—Juvenile literature. Earth (Planet)—Rotation—Juvenile literature. | LCGFT: Picture books. Classification: LCC QB521.5 .C3745 2019 | DDC 523.7—dc23
LC record available at https://lccn.loc.gov/2019024264

Word families:

"-ight"	→	br**ight**	n**ight**
"-out"	→	**out**	spr**out**
"-own"		d**own**	t**own**

Sight words:

and	at	day	each	help
is	me	not	the	there
up	where	you		

Bonus words:

across	dim	hiding	high	low
moves	sea	sky	warms	

Ready to go? Happy reading!

You might like to try answering the questions about the story on the last page of this book.

Where is the sun?

The sun is up high.

The sun is down low.

The sun moves
each day across
the sky.

The sun warms
the sea.

The sun warms
the town.

The sun warms you
and the sun warms me.

The sun is out.

The sun is hiding.

The sun and rain
help the plants sprout.

The sun is bright.

The sun is dim.

The sun is not out at night.

There is the sun!

Now that you have read the story, you might like to try to answer these questions.

1. As you go back through the story, can you count how many times you see the sun?

2. What do you like about daytime? What do you like about nighttime?

3. In this story, there are words that rhyme, such as "bright" and "night." Can you think of other words that rhyme with "bright" and "night"?

Great job!

You are a reading star!

YOUR SENSATIONAL SENSE OF
HEARING

BY JULIA VOGEL • ILLUSTRATED BY ROBERT SQUIER

The Child's World®

Published by The Child's World®
1980 Lookout Drive • Mankato, MN 56003-1705
800-599-READ • www.childsworld.com

ACKNOWLEDGMENTS
The Child's World®: Mary Berendes, Publishing Director
The Design Lab: Design and production
Red Line Editorial: Editorial direction

LIBRARY OF CONGRESS CATALOGING-IN-PUBLICATION DATA
Vogel, Julia.
Your sensational sense of hearing / by Julia Vogel ; illustrated by Robert Squier.
p. cm.
Includes bibliographical references and index.
ISBN 978-1-60954-286-3 (library bound : alk. paper)
1. Hearing–Juvenile literature. 2. Sound–Juvenile literature. 3. Ear–Juvenile literature. I. Squier, Robert, ill. II. Title.
QP462.2.V64 2011
612.8'5–dc22 2010037846

Printed in the United States of America
Mankato, MN
December, 2010
PA02068

ABOUT THE AUTHOR

Julia Vogel always has her ear to the ground for fun science facts. An award-winning author, she has a bachelor's degree in biology and a doctorate in forestry. Julia's all-time favorite sounds are wind in trees, rain, and children's outside voices.

ABOUT THE ILLUSTRATOR

Robert Squier has been drawing ever since he could hold a crayon. Today, instead of using crayons, he uses pencils, paint, and the computer. Robert lives in New Hampshire with his wife, Jessica, and a dog named "Q."

HI!

I'm Music Note. I love to listen to music. But I like other sounds, too. Have you heard crickets chirping on a summer night? What about thunder rumbling during a storm?

Your sense of hearing tunes you in to amazing sounds all around you. But what is a sense? And how does hearing work? Come along with me to explore your sense of hearing. Let the tour begin!

RAT-A-TAT!
RAT-A-TAT!

The rat-a-tat of drums fills the air. A colorful band marches past as you bite a hot dog. These sounds, sights, and tastes tell you a parade is passing by.

Hearing, seeing, and tasting are senses. Smelling and touching are, too. These five senses help you understand your world.

BOOM! BOOM!

Senses connect you to what's happening around you.

DING! DING!

Go catch the ice cream truck!

WOOF!

Look out for the angry dog.

Special body parts gather information about the world around you. Then they send it to your brain. Your eyes, nose, tongue, and skin are all **sense organs**. But my favorite sense organ lets you hear birds' songs and friends' voices. It's the awesome ear!

DING!
DING!
ICE CREAM
WOOF!

Ever wonder why you have two ears?

Ask a friend to pluck a guitar string by your left side. The sound is louder to your left ear. Having two ears lets you tell where the sound is coming from. If you were a rabbit, you could turn your ears to hear sounds better.

How does a guitar make a twang? Plucking a string makes it **vibrate**. It moves back and forth so fast, it blurs. The string's vibrations move the air, too. This makes the waves of sound. These **sound waves** ripple away from the string to your ear.

TWANG!

TWANG!

How do your ears work?

Your outer ears are more than places to hook your sunglasses. The ear flaps work like funnels. They gather sounds as soft as whispers. Want to hear even better? Cup your hand behind your ear. You'll hear everything your parents tell you!

The sound flows from your outer ear down into the dark opening you see. This is your **ear canal**. Hold tight! Our tour is headed inside your head!

Now we're in the middle ear. Here, sound waves hit a thin, tight covering. This is the **eardrum**. The eardrum vibrates, like the top of a drum.

The speed of the vibrations is important. Slow vibrations make low sounds. Fast vibrations make high sounds.

BOOM!

That's the slow vibration of a bass drum.

TWEET!

That's the fast vibration of a piccolo.

When the eardrum vibrates, three tiny bones linked to the eardrum move, too.

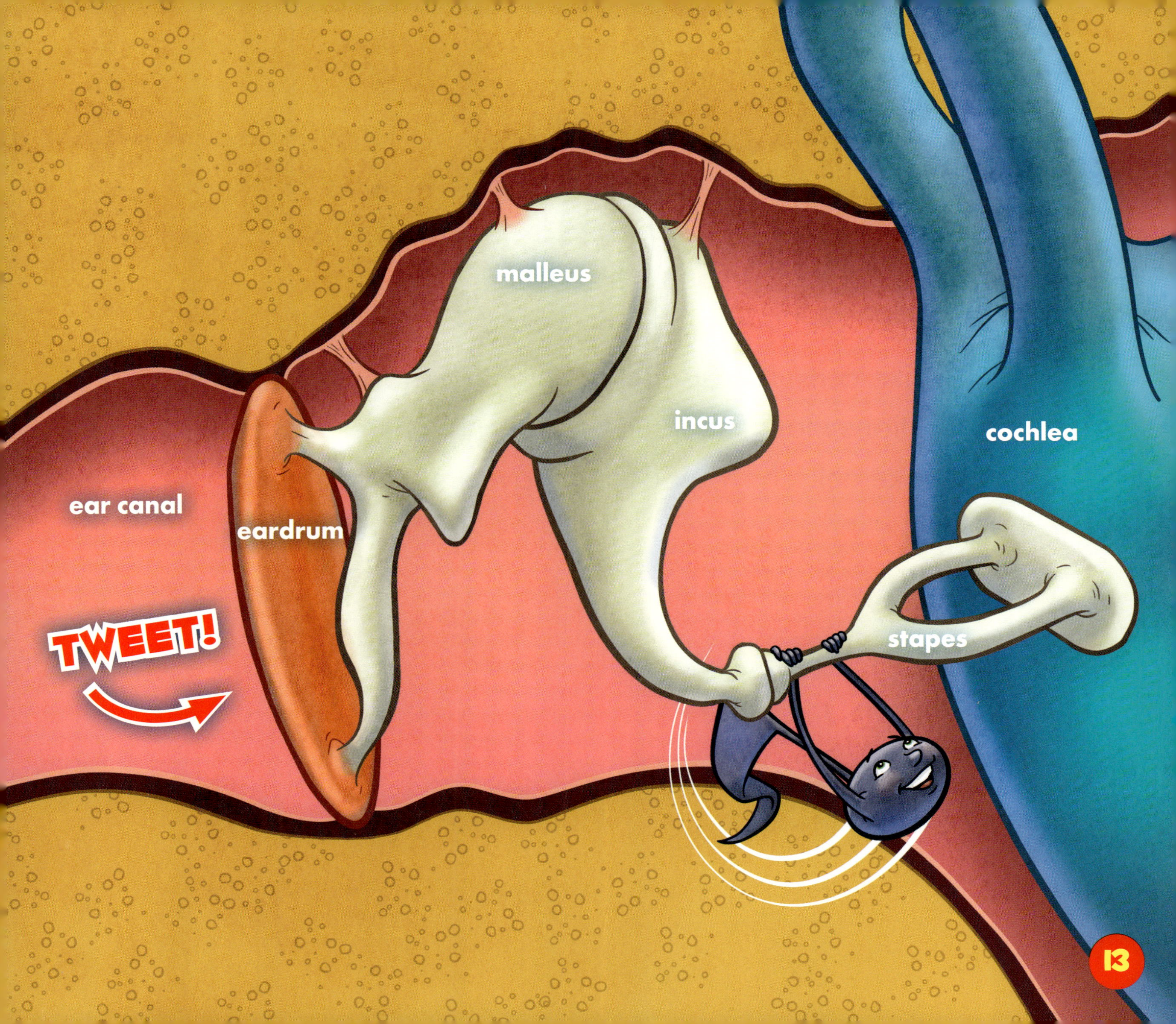
malleus
incus
cochlea
ear canal
eardrum
TWEET!
stapes

Now, in the inner ear, our tour takes some twists and turns. Part of it is shaped like a snail's shell. That's the **cochlea** (KOH-klee-uh).

The cochlea is lined with thousands of small hair-like cells. It is also filled with liquid. Sound vibrations make the liquid move. That moves the hair-like cells the way ocean waves move sea grass.

cochlea
stapes
hair-like cells

BAM!

Loud sounds move the hair-like cells more than soft sounds. Very loud sounds cause vibrations that toss the hair-like cells like a storm. This can damage the hair-like cells and harm your hearing. So keep the volume turned down, even for your favorite band!

But hearing is just getting started in your ears. In the inner ear, sound vibrations turn into messages. Those messages travel along **nerves** to your brain. The nerve pathways carry messages the way an Internet wire does.

The messages are **decoded** in your brain's special hearing centers. Your brain figures out if a sound is loud, soft, high, or low.

QUACK! QUACK!

Your brain lets you know it's a duck making that sound!

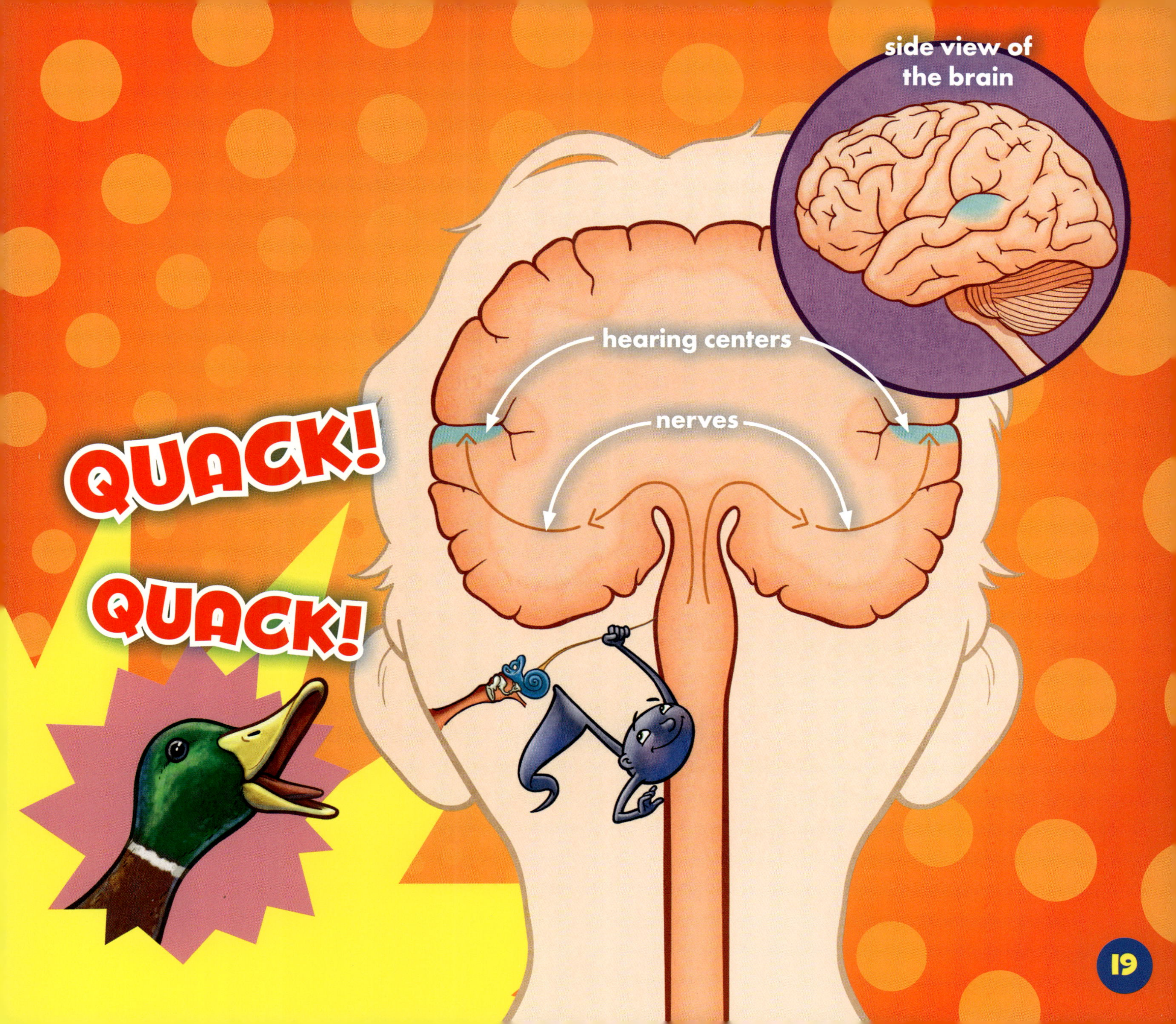
side view of the brain
hearing centers
nerves
QUACK!
QUACK!

WHOOSH!

Let's zoom ahead and explore your everyday life. How is hearing important to you?

DING, DONG! Some sounds give you good news. Your best friend is at the door.

PRRR. Your kitten likes when you pet her.

HONK! Other sounds are warnings. Step back! There's a car coming.

BUZZ. Careful! There's a wasp behind you.

Your brain listens for sounds of danger even while you sleep.

BUZZ
SODA
PRRR

Hearing also helps people understand each other, or **communicate**. They whisper, chat, yell, and–my favorite–sing!

WAAAH!

Babies tell their parents it's time for lunch. When they learn to talk, they can say, "More peas and carrots, please!"

Listen up! Hearing is a good way to learn. A teacher can help you solve a math problem or spell x-y-l-o-p-h-o-n-e.

Your hearing is usually best around age ten. As you get older, everyday noises can slowly add up—causing damage to your inner ears. Many people more than 60 years old have lost some of their hearing.

Doctors can fix some hearing problems with medicine or surgery. A **hearing aid** can make sounds louder. It acts like a tiny loudspeaker inside the ear.

There can be other bumps on the road to good hearing. Some babies are born with little or no hearing. An illness or head injury can also cause hearing loss. Someone who cannot hear is **deaf**.

People who are deaf can get help from trained dogs. Hearing dogs let their owners know if a smoke alarm is blaring.

American Sign Language is a special way to communicate without speaking. You can use your fingers to spell "Hi!"

HI!
HI!

Thanks for taking this tour with me. There is a lot more I'd love to show you about hearing.

We could head into the woods. There, we'd see how hearing helps animals find food and escape from enemies. Or we could dive into the sea. There you'd discover that sound travels faster in water than in air.

For now, remember that hearing helps you understand and enjoy the world. That's music to my ears!

HOOT!
HOOT!

LET'S JAM!

You can make your own musical instruments and listen for high and low sounds!

Things you need:

paper cups or cans
small pots or pans
rubber bands of different lengths and widths

Stretch the rubber bands around the different items. Be careful that you're not stretching one too tightly, or it could snap. Pluck the bands. Which ones sound higher or lower? Put a thick rubber band and a thin rubber band around the same item. Does one sound higher than the other?

Get some friends together and jam with your band.

GLOSSARY

cochlea (KOH-klee-uh): The cochlea is a part of the inner ear that is shaped like a snail's shell. The cochlea turns sound vibrations into messages for the brain.

communicate (kuh-MYOO-nuh-kayt): If people communicate, they share information, thoughts, ideas, or feelings with others. Hearing can help people communicate.

deaf (DEF): If someone is deaf, he or she cannot hear. Many people who are deaf use American Sign Language to communicate.

decoded (dee-KOHD-ed): Decoded means to have made a message or a signal understandable. Hearing messages are decoded in the brain.

ear canal (EER kuh-NAL): The ear canal is a passage where sound waves flow from the outside ear to the eardrum. The ear canal is in the outer ear.

eardrum (EER-drum): The eardrum is a thin, tight layer of tissue between the outer and the middle ear. The eardrum vibrates when sound waves hit it.

hearing aid (HEER-ing AYD): A hearing aid is a small device that a person wears in the ear to help him or her hear better. A person sometimes only needs to wear a hearing aid in one ear.

nerves (NURVS): Nerves are pathways that carry messages to or from the brain. Nerves carry messages about the things you hear.

sense organs (SENSS OR-gins): Sense organs are body parts such as ears, eyes, nose, tongue, and skin that help you understand your world. The ear is one of your sense organs.

sound waves (SOWND WAYVS): Sound waves are vibrations in the air that can be heard. Sound waves make the eardrum vibrate.

vibrate (VY-brayt): Vibrate means to move back and forth slightly and quickly. Sound waves make the eardrum vibrate.

FURTHER READING

Collins, Andrew. *See, Hear, Smell, Taste, and Touch: Using Your Five Senses*. Washington DC: National Geographic, 2006.

Furgang, Kathy. *My Ears*. New York: PowerKids Press, 2001.

Landau, Elaine. *The Sense of Hearing*. New York: Children's Press, 2009.

Walker, Richard. *Eyewitness Human Body*. New York: DK Publishing, 2009.

WEB SITES

Visit our Web site for links about your sensational sense of hearing:

childsworld.com/links

Note to Parents, Teachers, and Librarians: We routinely verify our Web links to make sure they are safe and active sites. So encourage your readers to check them out!